# Chapter 1

Facts of Hydrogen:

Hydrogen was discovered in the late 18th century by the English chemist and natural philosopher Henry Cavendish. Here's a brief overview of how hydrogen was discovered:

1.  Initial Observations: In the 17th century, Robert Boyle discovered that a flammable gas could be produced by the reaction of metal with an acid. This gas was initially referred to as "inflammable air."

2.  Discovery of Hydrogen by Cavendish: In 1766, Henry Cavendish repeated Boyle's experiments and found that the gas produced by the reaction of metal with acid was different from other known gases. He conducted a series of experiments to

determine the properties of this gas and concluded that it was a new element, which he called "inflammable air."

3. Further Studies: In the following years, Cavendish continued to study hydrogen and determined its atomic weight and density. He also found that hydrogen could react with oxygen to form water, a reaction that releases large amounts of energy.

4. Naming of Hydrogen: The name "hydrogen" was proposed by the French chemist Antoine Lavoisier, who recognized its unique properties and concluded that it was indeed a new element. The name "hydrogen" comes from the Greek words "hydro" (water) and "genes" (creator), reflecting its role in the formation of water.

Thus, Henry Cavendish discovered hydrogen through his repeated experiments and observations, and Antoine Lavoisier named it and recognized its unique properties, making it a new element in the periodic table.

Hydrogen is flammable because of its chemical properties and its reactivity with oxygen. Here's how:

1. Hydrogen is a highly reactive gas with a single electron in its outer shell. This makes it readily available to participate in chemical reactions with other elements, such as oxygen.

2. When hydrogen is mixed with oxygen, it can react rapidly to form water ($H_2O$) and release a large amount of energy in the form of heat and light. This reaction is highly exothermic and releases enough energy to ignite the mixture, causing it to burn.

3.   The flammability of hydrogen is further increased by its low ignition energy, which means that it only requires a small amount of energy to start burning. This makes hydrogen highly susceptible to ignition by sparks, heat, or other sources of ignition.

4.   The flammability of hydrogen is also influenced by its concentration in the air. At high concentrations, hydrogen is highly flammable and can pose a significant fire hazard. For this reason, hydrogen is stored and transported under carefully controlled conditions to minimise the risk of fire or explosion.

In conclusion, hydrogen is flammable because of its chemical reactivity with oxygen and its low ignition energy, which make it highly susceptible to combustion. These properties also make hydrogen a valuable fuel for many applications, but also require careful handling and storage to prevent accidents.

1. Hydrogen is the lightest and most abundant element in the universe, making up about 75% of its elemental mass.
2. It is a colourless, odourless, and tasteless gas that is highly flammable and explosive when mixed with oxygen.
3. Hydrogen has the chemical symbol H and is represented by the atomic number 1, making it the first element in the periodic table.
4. It has a single electron in its outer shell and is classified as a nonmetal.
5. In its elemental form, hydrogen is a diatomic molecule, meaning it is

composed of two hydrogen atoms bonded together ($H_2$).

6.   Hydrogen can be produced through several methods, including steam reforming of natural gas, electrolysis of water, and biomass gasification.

7.   It is used as a fuel in a variety of applications, including rocket propulsion, hydrogen fuel cell vehicles, and heating and cooling systems.

8.   Hydrogen is also an important raw material in the production of ammonia, which is used as a fertiliser, and methanol, which is used as a solvent and fuel.

9.   The use of hydrogen as a fuel source has gained popularity in recent years due to its ability to produce energy without producing harmful greenhouse gas emissions.

10.   However, the production and transportation of hydrogen still poses

challenges and requires further research and development in order to make it a more practical and efficient energy source.

Hydrogen is the most abundant element in the universe for several reasons:

1. Formation of Hydrogen: Hydrogen was formed in the early stages of the universe, through the process of Big Bang nucleosynthesis. During this process, the intense heat and pressure in the early universe caused hydrogen and helium to be formed from the primordial plasma.

2. Abundance of Hydrogen: Hydrogen is the lightest of all elements and is the most abundant element in the universe. This is due to its simple atomic structure, which consists of only one proton and one electron. This means that hydrogen atoms can be

easily formed and combined with other elements, making it one of the most abundant elements in the universe.

3.    Hydrogen in Stars: Hydrogen is also the primary fuel for stars, including our sun. Through the process of nuclear fusion, hydrogen atoms combine to form heavier elements, such as helium. This releases large amounts of energy, which powers the star.

4.    Hydrogen in Galaxies: Hydrogen is also found in vast quantities in galaxies, where it is primarily in the form of hydrogen gas. This hydrogen gas is a major component of the interstellar medium, which is the matter and radiation that exists between the stars in a galaxy.

Overall, the abundance of hydrogen in the universe is due to its simple atomic structure, its role in the formation of stars, and its presence in vast quantities in the interstellar medium of galaxies.

## Chapter 2

Storage and Containment of Hydrogen:

The storage and containment of hydrogen are critical to ensuring its safe use as a fuel and energy source. Hydrogen is a highly flammable gas and, when mixed with air, can pose a significant fire and explosion hazard. To minimise these risks, hydrogen must be stored and contained in such a way as to prevent the release of large quantities of gas into the environment.

There are several methods of storing hydrogen, each with its own advantages and disadvantages. One of the most common methods is compressed hydrogen storage, where hydrogen is stored in high-pressure tanks. Compressed hydrogen storage offers a relatively high energy density, making it suitable for applications where a large amount of energy is needed in a small space, such as in hydrogen fuel cell vehicles.

Another method of storing hydrogen is liquid hydrogen storage, where hydrogen is stored as a liquid at extremely low temperatures. Liquid hydrogen storage offers a much higher energy density than compressed hydrogen storage, making it suitable for applications that require a large amount of energy, such as in rocket propulsion. However, the low temperature required for liquid hydrogen storage makes it more difficult and expensive to handle and transport.

A third method of storing hydrogen is cryogenic hydrogen storage, which involves storing hydrogen as a supercritical fluid at very low temperatures and high pressures. Cryogenic hydrogen storage offers a high energy density and the potential for low-cost production, making it an attractive option for hydrogen storage.

In addition to the method of storage, the containment of hydrogen is also critical to ensuring its safe use. Hydrogen must be contained in such a way as to prevent its release into the environment, as well as to prevent the buildup of pressure that could lead to a fire or explosion.

One of the most common methods of hydrogen containment is the use of high-strength, lightweight materials, such as composite materials, to construct hydrogen storage tanks. These materials are able to withstand the high pressures required for hydrogen storage, while also being lightweight and easy to transport.

Another method of hydrogen containment is the use of permeation-resistant materials, such as metals and ceramics, to construct hydrogen storage tanks. These materials are able to prevent the release of hydrogen through permeation, ensuring that the hydrogen remains safely contained within the tank.

In addition to storage and containment, hydrogen must also be transported safely to ensure that it reaches its intended destination without posing a risk to the public or the environment. This is typically accomplished by using specialised hydrogen transport trailers, equipped with high-pressure hydrogen storage tanks, to transport hydrogen from the production facility to the end user.

Overall, the storage and containment of hydrogen are critical to ensuring its safe use as a fuel and energy source. By using high-strength, lightweight materials, permeation-resistant materials, and specialised hydrogen transport trailers, the risks associated with hydrogen storage and transport can be effectively managed, allowing hydrogen to be used safely and efficiently.

The materials used to store and transport hydrogen are critical to ensuring its safe use as a fuel and energy source. Hydrogen is a highly flammable gas, and it must be stored and transported in such a way as to prevent its release into the environment, as well as to prevent the buildup of pressure that could lead to a fire or explosion.

For storage, materials that are able to withstand the high pressures required for hydrogen storage and prevent the release of hydrogen are essential. The most common materials used for hydrogen storage are high-strength, lightweight materials, such as composite materials, that are able to withstand the high pressures required for hydrogen storage, while also being lightweight and easy to transport. These materials are typically made of carbon fibres embedded in a polymer matrix, and they offer a high strength-to-weight ratio, making them ideal for hydrogen storage.

Another type of material used for hydrogen storage is permeation-resistant materials, such as metals and ceramics. These materials are able to prevent the release of hydrogen through permeation, ensuring that the hydrogen remains safely contained within the storage tank. Common permeation-resistant materials include stainless steel, aluminium, and titanium, which are able to prevent the release of hydrogen due to their low permeability and high strength.

In addition to storage, materials that are able to withstand the high pressures and low temperatures required for hydrogen transport are also essential. The most common materials used for hydrogen transport are high-strength, lightweight metals, such as aluminium, titanium, and stainless steel, which are able to withstand the high pressures and low temperatures required for hydrogen transport, while also being lightweight and easy to handle. These materials are typically used to construct high-pressure hydrogen storage tanks, which are used to transport hydrogen from the production facility to the end user.

The materials used for hydrogen transport must also be able to withstand the corrosive effects of hydrogen, which can cause damage to the storage tank and lead to the release of hydrogen. To prevent corrosion, materials that are resistant to hydrogen corrosion, such as stainless steel, are commonly used for hydrogen transport. Additionally, coatings that are able to prevent the buildup of hydrogen on the surface of the storage tank, such as epoxy coatings, are also used to protect against hydrogen corrosion.

In conclusion, the materials used to store and transport hydrogen are critical to ensuring its safe use as a fuel and energy source. By using high-strength, lightweight materials, permeation-resistant materials, and materials that are resistant to hydrogen corrosion, the risks associated with hydrogen storage and transport can be effectively managed, allowing hydrogen to be used safely and efficiently.

Why is Hydrogen corrosive:

Hydrogen is considered corrosive due to its ability to react with certain materials, particularly metals. When hydrogen is stored or transported in metal containers, it can react with the metal and cause it to corrode, or deteriorate over time. This process of hydrogen corrosion is particularly damaging to metal surfaces, as it can create cracks, voids, and other defects in the metal that can weaken its structure and lead to the release of hydrogen.

The corrosiveness of hydrogen is due to its high reactivity and tendency to form hydrogen molecules that can penetrate into metal surfaces and cause chemical reactions. This process can cause the metal to become brittle and porous, which can weaken its structure and increase the risk of hydrogen leaks. In addition, the hydrogen molecules can also react with other substances, such as moisture, to form corrosive acids that can further damage the metal.

The risk of hydrogen corrosion can be reduced by using materials that are resistant to hydrogen corrosion, such as stainless steel, or by applying coatings that are able to prevent the buildup of hydrogen on metal surfaces. Additionally, the use of hydrogen purification techniques, such as drying and filtering, can also reduce the risk of hydrogen corrosion by removing contaminants and impurities from the hydrogen that can increase its corrosiveness.

In summary, hydrogen is considered corrosive due to its ability to react with certain materials, particularly metals, and cause them to corrode over time. To prevent hydrogen corrosion, materials that are resistant to hydrogen corrosion and coatings that can prevent the buildup of hydrogen on metal surfaces are used, as well as hydrogen purification techniques to remove contaminants and impurities that can increase its corrosiveness.

Chapter 3

Hydrogen Production Process:

Steam Reforming of Natural Gas

Steam reforming of natural gas is a process used to produce hydrogen from natural gas, which is a mixture of primarily methane and other hydrocarbons. This process involves the reaction of natural gas with steam to produce hydrogen gas and carbon dioxide. The steam reforming of natural gas is a crucial step in the production of hydrogen, and it is widely used as a source of hydrogen for various applications, including the production of ammonia, methanol, and other chemicals, as well as for hydrogen fuel cell vehicles. The steam reforming of natural gas involves two main reactions: the endothermic reaction and the exothermic reaction. In the endothermic reaction, natural gas is reacted with steam to produce carbon monoxide and hydrogen gas. This reaction is typically performed at high temperatures, between 700 and 900°C, and it requires an energy input to overcome the activation energy of the reaction.

The endothermic reaction can be represented by the following equation:

$$CH_4 + H_2O \rightarrow CO + 3H_2$$

In the exothermic reaction, the carbon monoxide produced in the endothermic reaction is reacted with steam to produce carbon dioxide and additional hydrogen gas. This reaction is performed at lower temperatures, between 400 and 700°C, and it releases energy in the form of heat.

The exothermic reaction can be

represented by the following equation: $CO + H_2O \rightarrow CO_2 + H_2$

The steam reforming of natural gas is typically performed in a reactor vessel, which is equipped with a heating system, a mixing system, and a cooling system. The natural gas and steam are mixed in the reactor vessel and then heated to the desired reaction temperature. The reaction products, including hydrogen and carbon dioxide, are then separated from the unreacted natural gas and steam and cooled to the desired temperature.

One of the benefits of steam reforming of natural gas is that it is a highly efficient process for producing hydrogen. The hydrogen yield from this process can be up to 95% depending on the conditions used and the purity of the natural gas. Additionally, the process is scalable and can be used to produce large quantities of hydrogen, making it ideal for industrial applications.

Another advantage of steam reforming of natural gas is that it is a relatively low-cost process compared to other hydrogen production methods. The cost of natural gas is relatively low compared to other hydrogen feedstocks, such as coal or biomass, and the process can be optimised to minimise the cost of hydrogen production.

Despite its benefits, steam reforming of natural gas also has some disadvantages. One of the main challenges of the process is that it produces carbon dioxide as a byproduct, which is a potent greenhouse gas. The carbon dioxide produced during the steam reforming of natural gas must be captured and stored, or it must be used for other applications, such as enhanced oil recovery.

Another challenge of steam reforming of natural gas is that it requires high temperatures to perform the reaction, which can increase the energy consumption of the process. Additionally, the process also requires a source of high-purity steam, which can increase the cost and complexity of the process.

In conclusion, steam reforming of natural gas is a process used to produce hydrogen from natural gas, which is a mixture of primarily methane and other hydrocarbons. This process is highly efficient, scalable, and low-cost compared to other hydrogen production methods, making it an ideal source of hydrogen for various applications. However, the process also has some disadvantages, including the production of carbon dioxide as a byproduct and the high energy consumption required to perform the reaction.

Electrolysis of Water:

Electrolysis of water is a process used to produce hydrogen and oxygen from water using an electrical current. This process involves the application of an electric current to water, which splits the water molecules into hydrogen and oxygen. The hydrogen and oxygen produced through electrolysis of water can be used for various applications, including fuel cells, hydrogen vehicles, and industrial processes.

The electrolysis of water is based on the principle of passing an electric current through a conducting solution, such as water, to produce a chemical reaction.

The process involves the use of two electrodes, typically made of metal, which are immersed in water. One electrode, the anode, is negatively charged, while the other electrode, the cathode, is positively charged.

When an electric current is applied to the electrodes, it causes the water molecules to ionise, or dissociate into positively charged hydrogen ions (H+) and negatively charged hydroxide ions (OH-). The hydrogen ions are attracted to the negatively charged anode, where they undergo oxidation, producing hydrogen gas. The reaction can be represented by the following equation: $H+ + e- \rightarrow H_2$

Similarly, the hydroxide ions are attracted to the positively charged cathode, where they undergo reduction, producing oxygen gas. The reaction can be represented by the following equation: OH- + e- → 1/2 O2

The hydrogen and oxygen produced during the electrolysis of water are typically separated from the water using a membrane or a separator. The hydrogen and oxygen produced can then be stored in containers for later use or further processed for specific applications.

One of the benefits of electrolysis of water is that it is a highly efficient and clean process for producing hydrogen. The process produces hydrogen with high purity, which makes it ideal for use in fuel cells and other applications where high-purity hydrogen is required. Additionally, the process produces hydrogen without emitting greenhouse gases, making it a clean source of energy.

Another advantage of electrolysis of water is that it can be performed using renewable energy sources, such as wind and solar power. By using renewable energy to power the process, the hydrogen produced through electrolysis of water can be considered a clean and sustainable energy source.

One of the challenges of electrolysis of water is that it requires a significant amount of electrical energy to perform the process. The energy required for electrolysis of water depends on the size of the system, the type of electrodes used, and the temperature and pressure of the water. Additionally, the process can be expensive due to the cost of the electrodes, the electrical energy required, and the need for specialised equipment.

Another challenge of electrolysis of water is that the process produces hydrogen and oxygen in equal amounts. This means that the hydrogen produced during the process must be stored, transported, and used efficiently to minimise the cost of the process. Additionally, the hydrogen produced through electrolysis of water must be used immediately, as it is highly reactive and can pose a safety risk if not handled properly.

In conclusion, electrolysis of water is a process used to produce hydrogen and oxygen from water using an electrical current. This process is highly efficient and clean, making it an ideal source of hydrogen for various applications. However, the process also has some challenges, including the high energy consumption required and the need for specialised equipment and expertise to perform the process safely and efficiently.

# Biomass Gasification:

Biomass gasification is a process that involves converting organic matter, such as wood, crops, and agricultural waste, into a mixture of gases, including carbon monoxide (CO), hydrogen (H2), and methane (CH4). The gases produced through biomass gasification can be used for various applications, including the production of heat and electricity, the generation of synthetic natural gas, and the production of fuels, chemicals, and fertilisers.

The process of biomass gasification involves heating the organic matter in a low-oxygen environment to produce the mixture of gases known as producer gas. The producer gas is then cleaned and conditioned to remove impurities and adjust the gas composition to meet the specific requirements of the intended application.

One of the main benefits of biomass gasification is that it is a highly efficient and sustainable energy source. The process utilises renewable organic matter, such as wood chips and agricultural waste, which are abundant and widely available. Additionally, the process produces a clean, low-carbon energy source that is less harmful to the environment than fossil fuels.

Another advantage of biomass gasification is that it can be performed on a small or large scale, making it suitable for both industrial and rural applications. The process can be used to generate heat and electricity for industrial processes, or to produce clean cooking fuel for households in rural areas.

Biomass gasification also offers a cost-effective and sustainable solution for managing agricultural and forestry waste. The process can convert waste materials, such as straw and wood chips, into useful energy and other valuable products, reducing the need for landfills and incineration.

One of the challenges of biomass gasification is that the process can be expensive and requires specialised equipment and expertise to perform effectively. Additionally, the process requires a continuous supply of feedstock, such as wood chips or agricultural waste, to maintain the energy production.

Another challenge of biomass gasification is that the gases produced by the process must be cleaned and conditioned to remove impurities, such as sulphur and particulates, and to adjust the gas composition to meet the specific requirements of the intended application. This cleaning and conditioning process can be expensive and requires specialised equipment and expertise to perform effectively.

In conclusion, biomass gasification is a process that involves converting organic matter into a mixture of gases, including carbon monoxide, hydrogen, and methane. The gases produced through biomass gasification can be used for various applications, including the production of heat and electricity, the generation of synthetic natural gas, and the production of fuels, chemicals, and fertilisers. Biomass gasification is a highly efficient and sustainable energy source that offers a cost-effective and sustainable solution for managing agricultural and forestry waste. However, the process can be expensive and requires specialised equipment and expertise to perform effectively.

# Photoelectrochemical Water Splitting:

Photoelectrochemical water splitting is a process that involves using light to split water into hydrogen and oxygen through the use of a semiconductor material. This process is considered a promising method for producing clean and renewable hydrogen, as it relies on the use of solar energy to power the reaction.

The basic principle of photoelectrochemical water splitting involves using a semiconductor material, such as silicon or a compound of titanium dioxide, that is capable of absorbing light and generating an electric current. This current is used to power the electrolysis of water, which splits the water molecules into hydrogen and oxygen.

One of the advantages of photoelectrochemical water splitting is that it allows for the production of hydrogen in a clean and sustainable manner. By using solar energy, the process avoids the need for fossil fuels and produces hydrogen with a low carbon footprint. Additionally, the process produces hydrogen and oxygen, both of which are valuable and versatile energy sources.

Another advantage of photoelectrochemical water splitting is its high efficiency. In laboratory settings, the process has been shown to have efficiencies of up to 10%, making it a promising method for large-scale hydrogen production. However, there are still technical challenges that need to be overcome to improve the efficiency of the process, such as developing more efficient semiconductor materials and improving the stability of the materials over time.

Despite its potential benefits, photoelectrochemical water splitting is still in the early stages of development for commercial use. One of the challenges of the process is that the materials used for the reaction can be expensive and difficult to produce on a large scale. Additionally, the materials can be sensitive to temperature and humidity, which can affect their performance and longevity. To overcome these challenges, researchers are actively working on developing new and improved materials for photoelectrochemical water splitting. This includes developing new types of semiconductors that are more efficient and stable, as well as improving the design of the photoelectrochemical cells to increase their efficiency and durability. Another challenge of photoelectrochemical water splitting is the need for high-intensity light to power the reaction. This means that the process is best suited for regions with high levels of solar radiation, such as desert regions or sunny climates. Additionally, the process requires large amounts of water to produce hydrogen, making it less suitable for regions with limited water resources. Despite these challenges, photoelectrochemical water splitting is considered a promising method for producing clean and renewable hydrogen. The process offers a number of benefits, including low carbon emissions, high efficiency, and the ability to produce hydrogen from a sustainable energy source. As the technology continues to develop and improve, it is likely that photoelectrochemical water splitting will become a more widely used method for producing hydrogen in the future.

Partial Oxidation of Hydrocarbons:

Partial oxidation of hydrocarbons is a process that involves using oxygen to partially oxidise a hydrocarbon-based fuel to produce hydrogen. This process is commonly used for the production of hydrogen from natural gas, and is considered to be a cost-effective method for producing hydrogen at a large scale. The basic principle of partial oxidation of hydrocarbons involves feeding a mixture of natural gas and oxygen into a reactor, where the mixture is subjected to high temperatures and pressures. The reaction between the natural gas and oxygen results in the formation of hydrogen, carbon monoxide, and other by-products. One of the advantages of partial oxidation of hydrocarbons is that it is a well-established process that has been used for many years in the chemical and petrochemical industries. This means that the process has a proven track record of safety and reliability, and the equipment used for the process is widely available and well-understood.

Another advantage of partial oxidation of hydrocarbons is that it is a cost-effective method for producing hydrogen. Natural gas is widely available and relatively inexpensive, making it a popular choice as a feedstock for hydrogen production. Additionally, the process is highly efficient, with yields of hydrogen ranging from 50-80%.

Despite its advantages, partial oxidation of hydrocarbons also has some disadvantages that need to be considered. One of the main disadvantages of the process is that it produces carbon monoxide as a by-product, which is a toxic and highly reactive gas. This means that the process requires careful management and handling to ensure the safety of workers and the environment.

Another disadvantage of partial oxidation of hydrocarbons is that it relies on the use of fossil fuels, which are non-renewable and contribute to climate change. While the process itself is relatively clean, the production of natural gas can be associated with greenhouse gas emissions, air and water pollution, and other environmental impacts.

To overcome these challenges, researchers are actively working on developing new and improved methods for partial oxidation of hydrocarbons. This includes developing new catalysts and reactor designs to increase the efficiency of the process and reduce the production of carbon monoxide. Additionally, researchers are exploring the use of renewable and sustainable feedstocks, such as biogas, to reduce the dependence on fossil fuels.

In conclusion, partial oxidation of hydrocarbons is a well-established and cost-effective method for producing hydrogen. However, it is not a perfect solution and has some disadvantages, including the production of carbon monoxide and the dependence on fossil fuels. To overcome these challenges, researchers are actively working on improving the process and exploring alternative methods for producing hydrogen.

## Fermentation:

Fermentation is a biological process that involves the conversion of organic compounds into energy or other useful products. It is a central metabolic pathway in many organisms and has been used for thousands of years for the production of food and beverage products, such as bread, beer, and wine. In recent years, fermentation has also been used as a method for producing hydrogen.

In the context of hydrogen production, fermentation involves using microorganisms, such as bacteria and yeast, to produce hydrogen through the conversion of organic compounds. The microorganisms consume sugars or other organic compounds, such as starch or cellulose, and produce hydrogen as a by-product of the metabolic process.

One of the advantages of using fermentation for hydrogen production is that it is a renewable and sustainable method. The feedstocks used for the process, such as sugars and starches, are derived from renewable sources, such as crops and waste materials, making it a more environmentally friendly option than the production of hydrogen from fossil fuels.

Another advantage of fermentation is that it is a simple and well-understood process. The technology required for the process is readily available and has been used for many years in the food and beverage industries. This means that the process has a proven track record of safety and reliability.

Despite its advantages, fermentation also has some disadvantages that need to be considered. One of the main disadvantages of the process is that it is relatively low-yielding, with yields of hydrogen ranging from 2-10%. This makes it less efficient than other methods for producing hydrogen, such as steam reforming or partial oxidation of hydrocarbons.

Another disadvantage of fermentation is that it requires the use of microorganisms, which can be expensive and difficult to manage. The microorganisms can be sensitive to changes in temperature, pH, and other environmental factors, making it challenging to maintain a stable and consistent process.

To overcome these challenges, researchers are actively working on developing new and improved methods for hydrogen production through fermentation. This includes developing new microorganisms with improved hydrogen-producing capabilities, improving the efficiency of the fermentation process, and reducing the costs associated with the process.

In conclusion, fermentation is a renewable and sustainable method for producing hydrogen. However, it is a relatively low-yielding and challenging process, with some disadvantages that need to be considered. Despite this, researchers are actively working on improving the process, making it a promising option for the future production of hydrogen.

Chapter 4

Uses of Hydrogen:

Hydrogen is a versatile and abundant element with a wide range of applications in various industries. In recent years, interest in hydrogen as a source of energy has increased, due to its potential as a clean and sustainable alternative to fossil fuels.

1. Energy generation: One of the most well-known applications of hydrogen is as a source of energy. When hydrogen is burned, it releases

energy in the form of heat and light, making it a potential fuel for power generation. Hydrogen can also be used in fuel cells, where it combines with oxygen to produce electricity and heat, with water as the only byproduct. This makes hydrogen a clean and sustainable alternative to traditional fossil fuels.

2. Transportation: Hydrogen is also used as a fuel for vehicles, including cars, buses, and trains. Hydrogen fuel cell vehicles use a fuel cell to convert hydrogen into electricity, which powers an electric motor. This makes them a clean alternative to traditional gasoline and diesel vehicles, as they emit only water and no harmful pollutants.

3. Chemical industry: Hydrogen is used in a variety of chemical reactions, including the production of ammonia, which is used in the manufacture of

fertilisers, and the production of methanol, which is used as a fuel and solvent. Hydrogen is also used in the refining of crude oil, where it is used to remove impurities such as sulphur and nitrogen.

4.   Space exploration: Hydrogen is used as a propellant in rocket engines, as it has a high energy content and a low molecular weight, making it an ideal fuel for space exploration.

5.   Refrigeration: Hydrogen is used as a refrigerant in some industrial cooling systems, as it has a low boiling point and is non-toxic and non-flammable.

6.   Metal fabrication: Hydrogen is used in the fabrication of metals, where it is used to reduce metal oxides to their respective metals. This process is known as hydrogen reduction, and is used in the production of iron, nickel, and other metals.

7. Medical applications: Hydrogen has been shown to have potential therapeutic benefits in medicine. In particular, hydrogen-rich water has been used in the treatment of various medical conditions, including oxidative stress, inflammation, and cancer.

8. Food and beverage industry: Hydrogen is used in the food and beverage industry, where it is used as a blowing agent in the production of carbonated drinks and as a reducing agent in the production of food additives.

9. Waste treatment: Hydrogen is also used in waste treatment, where it is used to break down organic waste and to produce biogas, a renewable source of energy.

In conclusion, hydrogen has a wide range of applications in various industries, including energy generation, transportation, chemical production, space exploration, refrigeration, metal fabrication, medicine, food and beverage production, and waste treatment. As interest in clean and sustainable energy sources continues to grow, the use of hydrogen is likely to become even more widespread in the future.

Fuel Cell:

A fuel cell is a device that converts chemical energy from a fuel source into electricity through an electrochemical reaction. Fuel cells have a number of advantages over traditional power sources such as batteries and combustion engines, including higher efficiency, longer life span, and lower emissions.

A typical fuel cell consists of two electrodes (anode and cathode) separated by an electrolyte. The anode is typically made of a metal catalyst, such as platinum, which is used to catalyse the reaction between hydrogen and oxygen. The cathode is also made of a metal catalyst, and is used to catalyse the reaction between hydrogen and oxygen on the opposite side of the electrolyte. When hydrogen is introduced into the anode, it splits into its constituent protons and electrons. The protons pass through the electrolyte and react with oxygen at the cathode to form water, while the electrons are forced to travel through an external circuit to reach the cathode. This flow of electrons through the external circuit generates an electric current, which can be used to power electrical devices. The fuel cell operates on a continuous basis as long as hydrogen and oxygen are supplied to the anode and cathode, respectively. The efficiency of the fuel cell is largely determined by the type of electrolyte used, as well as the temperature and pressure conditions in which the cell is operated.

There are several types of fuel cells, each with its own unique advantages and disadvantages. Some of the most commonly used fuel cells include proton exchange membrane fuel cells (PEMFCs), solid oxide fuel cells (SOFCs), and molten carbonate fuel cells (MCFCs).

PEMFCs are commonly used in applications such as automotive and portable power systems, due to their high power density, quick start-up time, and ability to operate at low temperatures. SOFCs are commonly used in large-scale stationary power systems, due to their high efficiency and long life span. MCFCs are commonly used in industrial applications, due to their ability to operate at high temperatures and their high efficiency.

Regardless of the type of fuel cell, all fuel cells have a number of advantages over traditional power sources. For example, fuel cells are highly efficient, with energy conversion efficiencies approaching 60-70%. They are also environmentally friendly, as they emit only water and heat as byproducts. Additionally, fuel cells can be designed to operate continuously for years without the need for maintenance or replacement, making them a cost-effective and reliable source of power.

Despite these advantages, there are also a number of challenges associated with fuel cells. For example, fuel cells require a constant supply of hydrogen and oxygen, which can be difficult to achieve in some applications. Additionally, fuel cells can be expensive to manufacture, due to the cost of the catalysts and other components.

In conclusion, a fuel cell is a device that converts chemical energy from a fuel source into electricity through an electrochemical reaction. Fuel cells offer a number of advantages over traditional power sources, including high efficiency, low emissions, and long life span. However, there are also challenges associated with fuel cells, such as the need for a constant supply of hydrogen and oxygen, and the high cost of manufacture. Despite these challenges, fuel cells are a promising technology that will likely play a major role in the future of energy generation.

Hydrogen in the Chemical Industry:

Hydrogen has a long history of use in the chemical industry, and is now widely used as a raw material for producing a wide range of chemicals, including ammonia, methanol, and others. In addition to its use as a raw material, hydrogen is also used as a source of energy for chemical reactions, as well as for hydrogenation processes that are used to modify the properties of certain chemicals.

One of the main uses of hydrogen in the chemical industry is in the production of ammonia. Ammonia is a key raw material for the production of fertilisers, and is also used in the production of nitric acid and other chemicals. In the production of ammonia, hydrogen is reacted with nitrogen to produce ammonia and water. This reaction is carried out in the presence of a catalyst, and requires high temperatures and pressures to occur.

Methanol is another important chemical that is produced using hydrogen. Methanol is used as a raw material in the production of formaldehyde, which is in turn used to produce a wide range of other chemicals, including resins and plastics. Methanol can be produced by hydrogenating carbon dioxide, or by reacting carbon monoxide with hydrogen in the presence of a catalyst.

In addition to its use as a raw material for chemical production, hydrogen is also used as a source of energy for chemical reactions. Hydrogen can be used to provide the energy needed to drive reactions such as oxidation, reduction, and hydrolysis. This is particularly useful in applications where traditional energy sources, such as heat, are not available or are not practical.

Hydrogenation processes are also widely used in the chemical industry, and involve adding hydrogen to various chemicals to modify their properties. For example, hydrogenation is used to produce a wide range of saturated fatty acids, which are used in the production of fats, oils, and other products. In addition, hydrogenation is used to produce a wide range of polymers, such as polyethylene and polypropylene.

Hydrogen is also used in the chemical industry for hydrogenation processes that are used to produce a wide range of specialty chemicals. For example, hydrogenation is used to produce a wide range of unsaturated fatty acids, which are used in the production of a wide range of specialty chemicals, including fragrances, flavours, and specialty polymers.

In addition to its use in the chemical industry, hydrogen is also used in a number of other industrial applications. For example, hydrogen is used as a coolant in many industrial processes, as it is highly effective at removing heat from various types of systems. Hydrogen is also used as a fuel for a variety of industrial processes, as it is a highly effective energy source.

Despite its widespread use in the chemical industry, there are also a number of challenges associated with hydrogen. For example, hydrogen is highly flammable, which can make it difficult to handle in certain applications. Additionally, hydrogen is highly reactive, which can lead to the formation of unwanted byproducts in some chemical reactions.

In conclusion, hydrogen is widely used in the chemical industry as a raw material, as a source of energy, and as a hydrogenating agent. Hydrogen is used in the production of a wide range of chemicals, including ammonia, methanol, and others, and is also used in a number of other industrial applications. Despite its widespread use, there are also a number of challenges associated with hydrogen, including its flammability and reactivity. However, these challenges are outweighed by the many benefits of hydrogen, and it is likely that hydrogen will continue to play an important role in the chemical industry for many years to come.

Space Exploration using Hydrogen:

Space exploration has been one of the most exciting and dynamic areas of human endeavour in the last century. The use of hydrogen in space exploration has been critical to the success of many missions, and has helped to push the boundaries of what is possible in this field. One of the primary uses of hydrogen in space exploration is as a fuel for rocket engines. Rockets rely on the rapid expansion of high-pressure gases to generate the thrust that is needed to escape the Earth's atmosphere and enter orbit. Hydrogen is an ideal fuel for rocket engines because it is lightweight and has a high energy content. When hydrogen is burned with oxygen, it produces a large amount of heat and high-pressure gas that can be used to drive rocket engines. Another important use of hydrogen in space exploration is as a coolant for electronic systems. In many space missions, the electronics must be kept cool in order to function properly. Hydrogen has excellent thermal conductivity, which makes it an ideal coolant for electronics in space. When hydrogen is used as a coolant, it can absorb a large amount of heat from electronic systems and dissipate it into space, helping to keep the systems cool and functioning properly.

In addition to its use as a fuel and coolant, hydrogen is also used in the production of life support systems for astronauts. In order to ensure that astronauts have a safe and reliable source of air and water, hydrogen is used to produce oxygen and water through a process known as electrolysis. In this process, hydrogen and oxygen are separated from water using an electric current, and the oxygen can then be used to support life.

Hydrogen is also used in space exploration to study the structure and composition of the universe. Hydrogen is one of the most abundant elements in the universe, and is a key component of many stars and galaxies. By studying the hydrogen in space, scientists can learn about the history and evolution of the universe, and gain insights into the processes that drive the formation of stars and galaxies.

Despite its many benefits, the use of hydrogen in space exploration is not without its challenges. For example, hydrogen is highly flammable, which makes it difficult to handle in certain situations. Additionally, hydrogen is difficult to store and transport, which can limit its use in certain missions. To overcome these challenges, scientists and engineers have developed a number of technologies and techniques that help to make the use of hydrogen in space safer and more efficient.

In conclusion, hydrogen has played a critical role in the success of many space missions, and has helped to push the boundaries of what is possible in this field. Whether as a fuel, coolant, or life support system, hydrogen has proven to be an indispensable resource for space exploration, and will likely continue to play a critical role in the future of space exploration.

Metal Fabrication:

Metal fabrication is the process of cutting, shaping, and assembling metal components to form structures, machines, and products. It is an important and widespread industry that plays a crucial role in many areas of modern society, including construction, manufacturing, transportation, and energy production. Metal fabrication can involve a wide range of processes, including cutting, welding, forging, stamping, bending, and assembly. The choice of process will depend on the type of metal being used, the desired product, and the required production volume. Some common examples of metal fabrication include the production of structural steel components for buildings, automotive parts, and machinery components.

One of the key benefits of metal fabrication is its versatility. Metal can be easily cut and shaped to meet specific design requirements, making it an ideal material for a wide range of applications. In addition, metal is strong, durable, and resistant to wear and tear, which makes it well-suited to demanding applications where reliability and performance are critical.

Another important aspect of metal fabrication is its role in promoting sustainability and reducing waste. Metal is a recyclable material that can be melted down and reused many times over, reducing the need for new materials and conserving natural resources. Additionally, metal fabrication techniques have become increasingly efficient and precise over time, reducing the amount of waste produced during the fabrication process and improving the sustainability of the industry.

However, metal fabrication can also have some negative impacts on the environment, such as the release of pollutants and greenhouse gases during the production of metal and the energy consumption of fabrication processes. To mitigate these impacts, many metal fabrication facilities have implemented sustainable practices, such as using renewable energy sources, reducing energy consumption, and reducing waste through recycling and reuse.

In conclusion, metal fabrication is an essential industry that plays a critical role in many areas of modern society. From the production of structural steel components and machinery parts, to the creation of sustainable products and practices, metal fabrication continues to be a key driver of technological progress and innovation.

## Medical Applications of Hydrogen:

Hydrogen has recently emerged as a promising new tool in the field of medicine, with potential applications in a wide range of areas, including cancer therapy, wound healing, and inflammation. One of the most exciting areas of research into hydrogen's medical applications is its potential as an antioxidant. Hydrogen is a highly reactive gas that has the ability to neutralise harmful free radicals in the body, which can cause oxidative stress and contribute to a number of chronic diseases, including cancer, cardiovascular disease, and diabetes.

In preclinical and animal studies, hydrogen has been shown to have potent antioxidant effects, and has been used to protect against radiation injury, reduce oxidative stress in brain and heart tissues, and improve outcomes in a variety of other diseases and conditions.

In addition to its antioxidant properties, hydrogen has also been explored as a potential therapy for cancer. In a number of studies, hydrogen has been shown to have anti-tumor effects, including the ability to inhibit the growth of cancer cells and promote cell death. Some researchers believe that these effects may be due to hydrogen's ability to modulate signalling pathways in cancer cells, leading to changes in gene expression and cellular behaviour.

Hydrogen has also shown promise as a wound healing agent, as it has been shown to promote tissue regeneration and reduce inflammation in preclinical studies. Some researchers believe that hydrogen's antioxidant and anti-inflammatory effects may be responsible for these benefits, as well as its ability to regulate cellular signalling pathways.

Furthermore, hydrogen has also been explored as a treatment for inflammation, which is a key contributor to a wide range of diseases, including rheumatoid arthritis, Crohn's disease, and ulcerative colitis. In a number of studies, hydrogen has been shown to have potent anti-inflammatory effects, and has been used to reduce inflammation and improve outcomes in animal models of these diseases.

Overall, the medical applications of hydrogen are still in the early stages of development, and much more research is needed to fully understand its potential and to bring these treatments to the clinic. However, the results so far are highly encouraging, and suggest that hydrogen has the potential to be a powerful new tool in the fight against a wide range of diseases and conditions.

# Hydrogen in the Food and Beverage Industry:

The food and beverage industry is one of the largest and most important sectors of the global economy, and it has been undergoing a transformation in recent years as new technologies and trends have emerged.

One of the most significant trends in the food and beverage industry is the shift towards healthier and more sustainable options. Consumers are increasingly seeking out foods and drinks that are free from artificial ingredients, preservatives, and chemicals, and are opting for products made from fresh, natural ingredients that are minimally processed. In response to this trend, many food and beverage companies have begun to focus on creating healthier, more sustainable products, using ingredients like whole grains, fruits, and vegetables, and avoiding artificial sweeteners and other chemicals. Some companies are even taking the concept of sustainability one step further by using renewable energy sources, reducing packaging waste, and using environmentally friendly practices in their manufacturing processes.

Another major trend in the food and beverage industry is the growth of specialty foods, such as gluten-free, organic, and vegan options. As consumers become more health-conscious, they are seeking out alternative options that cater to their specific dietary needs and preferences. The food and beverage industry has also seen a growing interest in artisanal and locally sourced products, as consumers look to support small, independent businesses and reduce their carbon footprint. Artisanal products, such as craft beers, artisanal cheeses, and specialty coffees, are becoming increasingly popular, and many consumers are looking for unique and locally sourced products that are made with care and attention to detail.

In addition to these trends, technology is also playing an increasingly important role in the food and beverage industry. With the rise of the Internet and e-commerce, many food and beverage companies are now able to reach new customers and markets, and are using technology to improve the supply chain and distribution processes, as well as to gather valuable data and insights about consumer behaviour.

Overall, the food and beverage industry is in a state of flux, with new trends, technologies, and consumer preferences constantly emerging and changing.

However, one thing is clear – consumers are increasingly seeking out healthier, more sustainable, and more unique products, and companies that are able to meet these demands will be well-positioned for success in the years to come.

Hydrogen uses in Waste Treatment:

Hydrogen is increasingly being used in various applications in the waste treatment sector, due to its ability to support various processes, including the reduction of greenhouse gas emissions, and the production of energy. The main use of hydrogen in waste treatment is in the process of waste-to-energy, where it is used to convert waste into a usable energy source.

One of the main ways that hydrogen is used in waste treatment is through the process of waste gasification. In this process, waste is heated in a controlled environment to create a gas that can be used as a fuel source. Hydrogen is produced as a by-product of this process, and can be captured and stored for later use. This not only reduces the amount of waste that is sent to landfills, but also generates a clean, renewable energy source.

Another way that hydrogen is used in waste treatment is through the process of anaerobic digestion. In this process, organic waste is broken down in an oxygen-free environment, with hydrogen being produced as a by-product. This hydrogen can then be used as a fuel source, and can also be converted into other forms of energy, such as electricity. Hydrogen is also being used in the treatment of wastewater, where it is used to help break down organic waste and reduce the levels of pollutants in the water. This process is known as the hydrogen-based advanced oxidation process (H2AOP), and it uses hydrogen peroxide and other hydrogen-based compounds to help remove pollutants from the water.

In addition to its use in waste treatment, hydrogen is also being used in the recycling of waste materials, such as plastics and metals. For example, hydrogen can be used to break down plastics into their constituent components, making it easier to recycle them and reducing the amount of waste that is sent to landfills.

Overall, the use of hydrogen in waste treatment has numerous benefits, including reducing greenhouse gas emissions, producing clean energy, and reducing the amount of waste that is sent to landfills. As the demand for more sustainable and environmentally-friendly solutions continues to grow, the use of hydrogen in the waste treatment sector is likely to increase in the coming years.

## Automobile Companies pursuing Hydrogen:

In recent years, many automobile companies have been pursuing the development and production of hydrogen-powered vehicles, in an effort to reduce greenhouse gas emissions and support the transition to a more sustainable energy system.

One of the key companies in this field is Toyota, which has been working on the development of hydrogen fuel cell vehicles (FCVs) for over two decades. The company's FCVs convert hydrogen into electricity, which is then used to power an electric motor, producing zero emissions in the process. Toyota's flagship FCV, the Mirai, is a prime example of the company's commitment to this technology, and has been well received by consumers.

Another major player in the hydrogen vehicle market is Hyundai, which has been working on the development of FCVs for over a decade. The company's first FCV, the ix35 Fuel Cell, was released in 2013, and since then Hyundai has continued to develop and refine this technology. The company's latest FCV, the NEXO, is one of the most advanced hydrogen-powered vehicles on the market, and has been praised for its performance, efficiency, and range.

Honda is another car manufacturer that has been pursuing hydrogen technology for some time, and the company has developed a range of FCVs, including the Clarity Fuel Cell. Honda is committed to the development of clean energy solutions, and has stated that FCVs have the potential to play a major role in the future of mobility.

In addition to these established automobile companies, a number of start-ups and new entrants are also entering the market, seeking to take advantage of the growing demand for hydrogen vehicles. These companies include Rivian, Fisker Inc., and Nikola, which are all developing hydrogen-powered vehicles with the aim of reducing emissions and promoting sustainability.

Overall, the pursuit of hydrogen technology by automobile companies is a promising sign for the future of sustainable mobility. As the demand for clean, efficient, and sustainable vehicles continues to grow, it is likely that more and more companies will invest in this technology, leading to further innovations and advancements in the field.

Countries using Hydrogen for Power Generation:

There are a number of countries around the world that are using hydrogen for power generation as part of their efforts to transition to a more sustainable energy system and reduce greenhouse gas emissions. Some of the leading countries in this field include:

1. Germany: Germany has been a pioneer in the use of hydrogen for power generation, and the country has a well-established hydrogen infrastructure, with a number of

hydrogen production and distribution facilities. In addition, Germany has set ambitious targets for the deployment of hydrogen technologies, and is working to support the growth of this sector through various initiatives and policies.

2.     Japan: Japan has been a major player in the development of hydrogen technology for many years, and has a well-established hydrogen fuel cell industry. The country is working to promote the use of hydrogen for power generation, transportation, and other applications, and has set ambitious targets for the deployment of this technology.

3.     South Korea: South Korea has been a leader in the development of hydrogen technology, and has a well-established hydrogen infrastructure, including production and distribution facilities. The country is working to

promote the use of hydrogen for power generation, transportation, and other applications, and has set ambitious targets for the deployment of this technology.

4. France: France has been working to promote the use of hydrogen for power generation, transportation, and other applications, and has set ambitious targets for the deployment of this technology. The country has a well-established hydrogen infrastructure, including production and distribution facilities, and is working to support the growth of this sector through various initiatives and policies.

5. China: China is one of the world's largest producers and consumers of hydrogen, and is working to promote the use of hydrogen for power generation, transportation, and other applications. The country has set

ambitious targets for the deployment of this technology, and is working to support the growth of this sector through various initiatives and policies. These countries, along with others, are working to promote the use of hydrogen for power generation as part of their efforts to transition to a more sustainable energy system and reduce greenhouse gas emissions. As the demand for clean, efficient, and sustainable energy sources continues to grow, it is likely that more and more countries will invest in this technology, leading to further innovations and advancements in the field.

www.ingramcontent.com/pod-product-compliance
Lightning Source LLC
Chambersburg PA
CBHW081711220526
45467CB00034B/2501